WILSON'S ORNITHOLOGY & BURDS IN SCOTS

Poems by
Hamish MacDonald

Illustrations by
Alexander Wilson

Introduction by
Paul Walton

Scotland Street Press
EDINBURGH

First Published in 2020
By Scotland Street Press
Edinburgh

This second edition published in 2022

Illustrations by Alexander Wilson
All illustrations with kind permission from the National Library of Scotland

Introduction by Paul Walton
© PaulWalton, 2020

ISBN 978-1-910895-39-9

Typeset and Design by Antonia Weir
Cover Design by Theodore Shack
Cover Artwork by Alexander Wilson

For

Kim, Kenna & Colla.

Contents

Introduction

ALEXANDER WILSON 1766-1813.

On a May afternoon in 1806, Alexander Wilson sat down on the wooded banks of the Kentucky river, took out his notebook and watch, and looked up. Overhead flew a flock of birds that was so immense, we struggle now to imagine the scene. The dense multitude flowed like a river in flood, moving in unison, so many birds that they obscured the spring sky. They were passenger pigeons and, less than a century after Wilson's death, the species had been driven to total extinction in the wild. The last passenger pigeon of all, Martha, died in a cage in Cincinnati Zoo in 1914.

Human actions have driven hundreds of species extinct since records began, and unquestionably many others before then. Each one has its own history of overexploitation, habitat loss, disruption, pollution, or struggle with novel diseases, competitors and predators introduced by people. Even in this litany, the story of the passenger pigeon is uniquely significant. We can say with confidence that this species was, before European colonists settled, the most numerous wild bird species in America and indeed was probably the most numerous wild bird ever to have existed on earth.[1]

This makes the passenger pigeon's extinction an incomparably powerful signal to humanity now, at the beginning of the Anthropocene epoch, as the full implications of human impacts on our environment begin to emerge around us.

We have Alexander Wilson to thank for the clarity of this signal. As he sat by the river that day, he did something that might now seem prosaic but which, at the time, was revolutionary. He set about systematically estimating the number of birds in the living torrent that passed above him, hour after hour. Not a wild guess driven to hyperbole, but a painstaking scientific assessment, based on direct observation and careful extrapolation. His conclusion was that this one passenger pigeon flock contained at least 2,230,272,000 – 2.2 billion - individual birds. Today, it is reckoned that the most numerous wild bird in the world is the red-billed quelea, a finch-like migratory African species that is a seed eater and a notorious agricultural problem. The quelea's entire global population is estimated at around 1.5 billion - just 2/3 of the size of the single passenger pigeon flock that Wilson witnessed that day.

---

1. The most common bird of all today is not a wild species, but the domestic chicken, estimated at 23 billion worldwide in 2017: three for every human being.

Wilson was not alone in writing about passenger pigeons. Tales of their vast flocks, of boughs breaking and falling from trees with the weight of nesting birds, of skies darkened and the thundering roar of wings, are common. So too are accounts of the pigeon hunting industry that sprang-up in the mid-19th century, and its methods. Pigeons were shot and trapped with nets, torched at their roosts, asphyxiated with burning sulphur, blown-up with dynamite. They were attacked with rakes, pitchforks, poisoned with whiskey-soaked corn. There is no shortage of vivid description in the surviving accounts of this bird, and its demise.

The difference with Wilson's account is that his is an objective and systematic estimate of a population. We know we can rely on it because his descriptions, paintings, impressions and analyses of other birds, ones still around now, are astonishingly accurate. His passenger pigeon estimates can only be correct.

Those impressions of birds are in Wilson's key work, his nine-volume American Ornithology. It describes 268 species in detail. He drew and painted these birds, often – again ground-breaking at the time – directly from life, setting them imaginatively but accurately in their habitats, colouring plumage details with unprecedented precision. The illustrations were accompanied by text that was engaging, informative, honest, and inclusive, intended to be read by everyone and to be enjoyed. The work contains jokes. Yet American Ornithology was easily the most advanced and comprehensive academic ornithological work of its time.

As a scientific contribution, produced as modern biology

was in gestation for its late 19th century flourishing, Wilson's work is of lasting academic significance. It is, simultaneously, an artistic and literary triumph. For all this, Wilson is renowned across the USA, his name forever synonymous with the study, appreciation and admiration of birds, peppering the natural history lexicon of America: Wilson's warbler, Wilson's snipe, Wilson's petrel, Wilson's phalarope; the Wilson Ornithological Society, the Wilson Bulletin of Ornithology.

He remains, however, curiously obscure and unappreciated in the Old World, and Scotland in particular. Curious because American Ornithology, this dominant landmark of natural history, was written and illustrated by a small-time salesman, journeyman weaver, self-taught poet and working-class radical from Paisley.

In this volume, Hamish MacDonald brings Wilson's art together with contemporary wit and ornithological wisdom in the Scots dialect that Wilson himself spoke and wrote. Hamish does us a grand and overdue service: here we begin to refocus Scots onto the singular Alexander Wilson legacy – an' we hae warsels a braw time alangside.

Wilson was born in 1766, the son of 'Saunders' Wilson, a former soldier, weaver and whisky distiller. Whether the whisky was legal or otherwise is unconfirmed. Given that his mother Mary McNabb was from the village of Rhu on the east shore of Gare Loch in Argyll, however, the prospect of illicit distillation looks certain. Illegal production and movement of the spirit was common at this time, and it was not always clamped down on by Scots officials with the vigour that Southern authorities might have expected. Rhu was a major centre of illicit whisky production. Aldownick Glen, a ravine close to the town, held several illegal distilleries. In the Heart of Midlothian, Walter Scott describes whisky smuggling around Rhu, and to this gorge specifically, under the name 'Whistlers' Glen'. Lookouts gave warning of the approach of unwelcome visitors by imitating the call of the curlew. The area was renowned. King George IV visited Scotland and raised eyebrows by asking for a taste of 'real smuggled whisky'. The Duke of Argyll had to undergo the discomfort of meeting the smugglers face-to-face on the exposed beach at Rhu Point to procure for his monarch a barrel from the Whistlers' Glen.

The Wilson family income was enough for a house by the White Cart river where it runs through Paisley town. Alexander (Sannie to his folks) was born here. Testament to the longstanding connections between Scotland and North America - still manifest today in Donald Trump's immigrant roots from the Isle of Lewis - Alexander Wilson was baptised in Paisley by Rev Dr John Wetherspoon who, ten years later, left Scotland and eventually became a signatory to the Declaration of Independence.

Tall, dark-haired young Wilson played along the White Cart and attended Paisley Grammar. A career in the Presbyterian ministry was Mary McNabb's plan, with private tutoring hired-in to that end. When he was ten, however, his mother died, his father remarried a widow with children of her own, and Wilson's formal education ended abruptly. The family of eight moved out to part-ruined Auchenbathie Tower near Lochwinnoch – reputedly the ancestral home of William Wallace's family and defended by him and his men against English raiders in the 13th century. With the Tower walls right for stills, and the contraband hub of Beith nearby (the pub on Beith Mainstreet is still the Smuggler's Arms), Auchenbathie Tower well suited his father's activities in the drinks industry.

Alexander supplemented the family income by tending cattle on the common grazings in the hills and woods in this area, and it is here, aged twelve, that he began to write poetry:

> Castle Semple stands sae sweet,
> The parks around are bonnie, O;
> The ewes and lambs ye'll hear them bleat
> And the herd's name is Johnnie, O.

In his early teens, he was back working in Paisley, apprenticed as a weaver under his brother-in-law William Duncan. William became alcoholic and difficult - but Wilson grew close to his nephews and nieces, especially the eldest William - the two were to emigrate together, and to be friends for life. Wilson learned music and dance during this time, he wrote more poetry in Scots dialect, and his interest in nature began to intensify. He acquired a rifle and became, by contemporary accounts, an incredible marksman, poaching game for the family table, and more: he used the gun to develop an intimate knowledge of the wild birds around him. Before the advent of light, affordable optics, the nuances of bird identification and taxonomy required some shooting.

On completing the apprenticeship Wilson worked at the weaving in Lochwinnoch and later in larger loom shops in Paisley. He developed a reputation for inventing funny verse while working at the loom. When in 1786 Burns published his first volume, Poems, Chiefly in the Scottish Dialect, the potential for expression through, and recognition for, literary efforts in native Scots underwent a sudden step-change, and ambition to be a poet grew in Wilson alongside. His horizons were also expanding as he spent less time at the loom and more as travelling salesman, moving throughout the Scottish mainland on foot, selling cloth and silk - and writing.

> The morn was keekin' frae the east,
> The lav'rocks shrill, wi' dewy breast,
> Were tow'ring past my ken,
> Alang a burnie's flow'ry side

An itinerant life interspersed with periods working at the looms suited him, but poetic ambitions ran high. Aiming for publication, he struck deals with Paisley printers: if whilst travelling he could sell, alongside the cloth, an agreed number of subscriptions to a volume of poetry, publication would proceed. This mechanism left him in debt, but it worked. His first book Poems appeared in 1790.

With that, the 24-year-old entered the literati of the Scottish enlightenment. He consolidated by travelling from Paisley to Edinburgh, on foot, to debate the merits of Scots poets at the Pantheon Club before of an audience of over 500 capital worthies. A second edition followed, published in Edinburgh, publications in literary magazines and, inspired by Burns' 1791 triumph Tam o'Shanter, he anonymously published Wattie and Meg, an adaptation of the Taming of the Shrew in verse.

It is striking that access to the higher strata of society was, in late 18th century Scotland, effectively open to all comers with something to say, and the will and energy to share it. That impression perhaps speaks more to the exclusivity of cultural establishments today, than the exceptions of the time. Equally striking is the political groundswell underway around young Wilson as the 1790s began. He was soon finding things to say in this regard too, things that would prove pivotal in his future. In that year he published The Hollander, or Light Weight, a satire describing the atrocious working conditions in the Paisley textile mills, and the blatantly exploitative activities of many of their owners. As a plea for unionisation, it is among the earliest protest literature of the industrial revolution. It made Wilson an admired figure among the workers and contemporary reformers – including writers, artists and musicians in the Friends of Reform group. It also led to exploratory, for now curtailed, libel action against Wilson from the slighted mill owners.

In May 1792, conflict escalated. William Sharp, owner of the Long Mills, received an anonymous letter, apparently in Wilson's hand (he always denied authorship) seeking to extort £5 in return for suppressing publication of a new poem, The Shark, which was certainly by Wilson, and which told how mill owner 'Willy Shark' cheated his brutally oppressed workers by using a doctored cloth-measuring rule to assess their output. Sharp immediately petitioned for Wilson's arrest. He was imprisoned in Paisley Tollbooth for a month, and later fined. Three further spells in prison followed, owing to his inability to meet fines and costs, and some more inflammatory Reformer prose. In February 1793 the court instructed him to destroy his writings openly in the Paisley square, which he did, by fire.

With France in the grip of The Terror, reformers found guilty of treason or sedition faced transportation or the death penalty. Wilson enjoyed growing renown among the working classes and in his case, the authorities appeared mindful of creating a martyr among the Reformer movement. Wilson was freed, but the libel and blackmail charges were never dropped. He lived in these febrile times at constant risk of arrest, incarceration, or worse.

The appeal of a new land with a new culture of equal opportunity - one where a criminal record in Scotland would effectively be erased - is obvious. In May 1794, 28-year-old Wilson and his 16-year-old nephew William Duncan walked to Port Patrick, caught a boat to Belfast, and there they secured deck-space on the Swift, bound for Philadelphia. He never returned to Scotland.

During his first years in America Wilson worked as an engraver, weaver and travelling salesman, and latterly as a school teacher. Political sensibility re-emerged and induced a return to writing. In the election year 1800, his poem Jefferson and Liberty became partisan campaign literature for the candidate, and it shaped a friendly relationship between Wilson and the President following his inauguration in 1801.[2]

2. Jefferson purchased one of the first copies of American Ornithology, following a sales visit by Wilson to the White House. The copy remains in the Library of Congress.

That year was a key moment for Wilson. He took a teaching post at the Union School at Kingsessing, Pennsylvania. There is nothing extraordinary in that - other than this school being close to the house and garden of the Bartram family. William Bartram was a botanist, traveller and philosopher.

He was the son of John, a botanist and horticulturalist, and correspondent of Carl Linnaeus – who pronounced him 'the greatest natural botanist in the world'. John had died in 1777, but had established a botanical garden, which William now maintained. William was the foremost naturalist in America at the time, had an extensive library, travelled widely, was himself an industrious recorder of birds, and he and Wilson became close friends.

The meeting drew Wilson into a world of natural history, of observation and recording, drawing and painting, of organising and communicating these records. It was a new use for his skills and a new purpose. The 'great naturalist' John Bartram had been self-made, and almost entirely self-educated. For Wilson this was American opportunity and promise made manifest. Scientifically, the Linnaeus connection was critical. The Swede had invented the binomial system of naming species – as in Homo sapiens – which proved to be a key breakthrough in the classification of life on Earth. The system was not the only one at the time, but history would prove it to be easily the best and most useful. It was the natural approach for William Bartram, and now for Wilson too. Through the lens of this system, the disarray and confusion surrounding the birds of America was obvious to the two naturalists. At this time, many eminent

European naturalists characterised American birds as derivative of European archetypes, as secondary to their European relatives, and somehow subordinate to them. The confused and gap-riddled nature of American bird classification and cataloguing underscored this absurd notion.

Wilson took it upon himself to put it right. He donned buckskins, took up his rifle, paints and notebooks, and set off, on foot, from Pennsylvania to Niagara, returning by an alternative route. It was the first of many long field expeditions across the continent. Wilson is estimated to have walked at least 12,000 miles between 1804 and 1813, recording birds all the way, visiting every state on the US mainland and developing a comprehensive overview of the country's bird fauna. He also built a wide network of friends and contacts, whom would become correspondents while he was back writing and illustrating in Pennsylvania, passing on the local knowledge in which American Ornithology abounds. His specific vision for the work seems to have been clear from the outset. He took great pains over bindings, papers, print quality and colours, and first editions are objects of rare elegance and beauty. Utility was central, however: one is struck by how similar the layout and content of American Ornithology is to a modern-day bird field-guide. This was, foremost, a vehicle for sharing understanding and enjoyment of the living world.

These long solo scientific expeditions were also sales trips, just as he had made across Scotland selling cloth in his late teens. His funding mechanism for American Ornithology involved building a bank of subscribers, each making a down-payment for the next volumes whilst they were still in production. Moreover, much of this travel will have been through remote and wild lands only newly, if ever, visited and touched by Europeans: when James Fenimore Cooper was interviewed following the success of The Last of the Mohicans, he was asked if the character Hawkeye was inspired by any real person. He replied, 'Based him on a Scotchman called Wilson'. Whether this was Alexander is not confirmed, but it is undeniable that the label 'adventurer', alongside poet, painter, naturalist, salesman, radical – all these cleave quite naturally to Wilson, who nevertheless remained, by all accounts, self-effacing, modest and humorous company.

The outstanding feature of Wilson's work, however, is the easy, effortless, instinctive marriage of art and science that is manifest in American Ornithology. The contrast with the interdisciplinary ravines and chasms that punctuate the intellectual and cultural landscape today, is palpable. It feels like a loss, and a significant one. It looks to have been a product partly of the man, but also of the times. By 1865, Walt Whitman was writing:

> When I sitting heard the astronomer where he
> lectured with much applause in the lecture-room,
> How soon unaccountable I became tired and sick,
> Till rising and gliding out I wander'd off by myself,
> In the mystical moist night-air, and from time to time,
> Look'd up in perfect silence at the stars.

That sentiment resonates today – and whilst it has a modern gravity, the notion that investigation, measurement and the quest for understanding must necessarily kill beauty, must undermine spiritual responses to the natural world, would surely have been alien to Wilson just half a century earlier. How obvious is it now that he and his contemporaries were richer for that innocence.

In the end, Wilson's was perhaps the most perfect American Dream. Child of the Scottish Enlightenment, driven by persecution to America, where he was free to roam and to think, to earn through mind and muscle, talent and drive, to pursue his own unique happiness and build a profound legacy along the way. Wilson died in Philadelphia in 1813 aged just 47. Over-work to complete American Ornithology appears to have been the central contributor to his failing health. He died proudly, legally, emphatically an American. But he was born and raised, equally emphatically, a Scot, and in the pages that follow Hamish MacDonald, at long last, reaffirms and celebrates this truth.

Paul Walton
Head of Species and Habitats for RSPB Scotland
Glasgow
January 2020

WILSON'S ORNITHOLOGY
& BURDS IN SCOTS

Poems by
Hamish MacDonald

Illustrations by
Alexander Wilson

Scotland Street Press
EDINBURGH

THE SPECHT / WOODPECKER

Wi a dirl an a drum on the tapmaist tree
Wi a rattle an a dirl an a drum
Chappin awake the springtime day
Wi a rattle an a dirl an a drum

Drillin doon fir the eemock an the wirm
Wi a rattle an a dirl an a drum
Scran tae feast on frae the bark an the beuch
Wi a rattle an a dirl an a drum

Dirrin on the trunk fir a mate's call back
Wi a rattle an a dirl an a drum
Borin oot a hole fir the scawdie's nest
Wi a rattle an a dirl an a drum

Pileated woodpecker
Ivory Billed Woodpecker
Red headed Woodpecker

A scarlet blent on a bleck/white plume
Wi a rattle an a dirl an a drum
A wing-spreid fan when he taks tae the lift
Wi a rattle an a dirl an a drum

This is the specht, the spottit widpecker
Wi a rattle an a dirl an a drum
The rat-a-tat avian Black an Decker
Wi a rattle an a dirl an a drum

GUOMUNDUR THE STIBBLE-GUISS /
GUOMUNDUR THE GREYLAG

Guomundur flew frae Iceland
In ae foulakauvie o five hunder geese
Tae feed on Gruinart's stibble
An fin some winter peace.
But he's quite content tae bide a while
As wintry days are coolin
For he's caught a whuff o the Angel's Share
Frae the stills o Lagavoulin.
For he's caught a whuff o the Angel's Share
In coorse Atlantic weather
An dreams o hame an simmer dim
Amang the Arctic heather.

Goosander Duck & female
Pin-tail Duck
Blue-wing Fowl
Snowgoose

4

THE CORBIE / CROW

The corbie staunds upon a stane
This warld tae determine
Tae craw some judgement frae his beak
Ae fire an brunstane sermon.

Whan corbies meet in catticlour
Wi racous clishmaclavers
Wha kens whit fash or duelsome waes
Maun gar sic antrin haivers.

For they've aye been kent tae tak the air
In coal-black murmurations
Tae cairry coffins on ther backs
Tae unseen destinations.

An they hae ae word fir Corbie meets
It's kent tae scholart an tae burder
Whaure'er they've gaithert in a flock
They'll say "Ther's been a murder".

Crow, Touche Bunting

THE SMAW DOUKER / LITTLE GREBE

The Smaw Douker bides amang the reeds
biggs her nest frae the wattery weeds
babs an jouks whan a mate cams cawin
pleeps an treels in the day that's dawin
douks aneth the lochan fir the mealtime catch
watter-golach, beardie an the mayflea hatch
scuddin doon throu benmaist neuks
ther's nane can douk brawer than the Smaw Douker douks.

Eider Duck & Female, Snow Duck, Ruddy Duck & Female

Red Bellied Woodpecker, Cedar bird, Yellow throated Flycatcher, Purple Finch

WEXWINGS / WAXWINGS

Ae winter the wexwings came
tae gift-wrap the gairden tree
flame crestit wi goth eye-liner
bauld splatches o colour frae the penter's brush
intruded in a snaw-white warld.

Icicles baured us in oor winter prison
as we cooried in
agin the Arctic cauld.

Days later A luiked oot
but the vanishin act
wis abrupt as thon first appearance.

Wi each passin winter A'd watch
bidin ma time fir the riot o colour
but ther wis only smirr an deid air
a hingin waff o muilderin decay
as lang months passed in daurkness.

Still.
A'll wait fir the waxwings again this winter
nature's lottery ticket
turnin up agin the odds.

Great White Heron, Night Heron & Young Night Heron

SWAN HAIKU

swan on Hielan loch
sail on by wi galleon grace
turnin day frae nicht

HOULET / OWL

Houlet, houlet fly by nicht
Poised for faintest soond or sicht
By munelit knowe or shadit neuks
Wi clauts as shairp as fishin hooks
Skimmer past on silent flicht
Houlet, houlet fly by nicht.

Barred Owl, Short Eared Owl

DORIC BIRDSANG

Fit-like-min?

    Fit-like-fit-like-fit-like?

        Tyaavin-awa-Tyaavin-awa.

Foo's it gan?   Foo's it gan?

Yer seein it.

Yer seein it.

Maryland yellow throat, Yellow breasted Chat, Female red bird, Indigo bird, American red start

THE BRONGIE /
CORMORANT

The Brongie dawdles on the rock
abune the cauld grey sea
whiles claikin tales o wracks an gales
wi a glent in its sapphire ee.

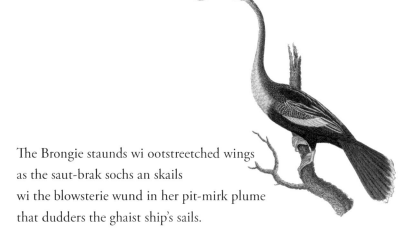

The Brongie staunds wi ootstreetched wings
as the saut-brak sochs an skails
wi the blowsterie wund in her pit-mirk plume
that dudders the ghaist ship's sails.

The Brongie loups frae stob an stane
whaur the lanely bell buoy tolls
tae sowm an sweep through the briny deep
in the reive o the siller shoals

Then Brongie sprachles hamewart
wi its thrapple stappit fu
fir the younkers tae dine in white dickies so fine
at The Cormorant's Rendezvous.

Black bellied Darter, Great northern Diver, Black headed Gull, Little Auk, Female Darter

BIRDSANG O THE WEST

See-you-pal. See-you-pal.
Giez a brekk. Giez a brekk. Giez a brekk.
Squerr-go-then. Squerr-go.
Bolt-ya-rocket. Bolt-ya-rocket.
Ya-brammer-ya-brammer-ya-brammer.
Ferr-dooz.
Ferr-dooz.
Ferr-dooz.

Greefer, Golden crested Wren, House Wren, Black capt Titmouse, Crested Titmouse, Winter Wren

THE GOWK / CUCKOO

Dae ye hear yon gowk
in the alder den
whan the April souch cams blawin
an the bog-spink flouers
in the mossy glen
while the snaw-wreath's saftly thawin.

An it's cuckoo here an it's cuckoo ther
frae the dernie howe cams ringin
but ther's scant a sicht o the cuckoo burd
while the joukie craitur's singin.

Begowk us aw
ye sleekit wratch
frae in yon leafy border
ye twafauld, skellum, swickit burd
an rogue o the heichmaist order!

For they'll drap ther eggs
in the dunnock's nest
the rabbin or the pipit
an syne they'll cowp ilk ane fir ane
till twal or mair are slippit.

Sae fausely-bred ya fledgeling ned
ye've tore the clatch asunner
this blaw-in chancer coories in
whiles taks tae deith an plunner.

An it's cuckoo this an it's cuckoo that
frae the dernie howe a-ringin
wi a robber's ee on the greenwid tree
whan yon joukie craitur's singin.

Black Billed Cuckoo, Yellow Redpoll

PYOT JEWEL THIEF (ALLEDGEDLY) / MAGPIE THE JEWEL THIEF (ALLEDGEDLY)

The Pyot is a gear-gaitherer
She loos the bonnie trinkets
Tae stash in nests an hidie-holes
Whaure'er they micht be slinkit
The geegaw or the gemstane
The precious wallie-gowdie
The diamant an the sovvie-ring
The troke an whigmalorum.
For ilka objeck that she'll huird
Wi glent or brinnin twinkle
She pits the ee-gless tae her brou
Tae swatch the bonnie skinkle.

The Magpie

Warblers & Humming birds

Yellow Crowned Heron, Great Heron, American Bittern, Least Bittern

HERON

The Heron stalks the lanely streams
Feedin on minnas an fisherman's dreams
Stieve as a statue
Seelent and slee
He'll ding wi his spear in the blink o an ee

As wide-eened puddock pechs for a braith
Then oars tae the deep tae avoid certain daith
Through fairy-moss tendril
An mirkie lagoon
Jinkin the dabs o the whaler's harpoon

The heron flees hame wi a richt puggelt straik
High up in the branches tae skrauch an tae skraik
In the heronry clachan
Crack news o the day
An tell tales o yins that got away

AILSA PAURIT/THE PUFFIN

The Ailsa Paurit gairds the Craig
in snaw-white downie tabart
a pirate coat upon his back
an cutlass in his scabbart.

Huffin an puffin amang the gress
it widdles then it scurries
wi colourt neb tae brod an jeb
the seamaw frae the burraes.

High on the sea-girt curlin stane
Wi cruin an gurly greetin
Feedin on the Tammie-yaw
Whan simmer tides are fleetin.

Great Tern, Stormy Petrel

THE CROSSBILL

The crossbill is a bonny burd
An she sings wi a guid Scots tongue
Jip jip jip
A'll gie ye gip
Gin ye meddle wi me nor ma young

The crossbill is a brawlike bird
An she dines on the cones an the nits
Her neb is unique
Wi a crossower cleek
An the heich pine croon's whaur she flits

The crossbill's a hamefarin bird
An she trills her dowie sang
By the hilltaps o Straloch
Or by wild foamin Falloch
Contentit tae bide saison lang

American Crossbill, Female, White-winged
Crossbill, White-crowned Bunting

The Mississippi Kite

THE HAME-COMIN O THE RID GLED /

THE REINTRODUCTION OF THE RED KITE

It's been auld lang syne sin Rid-gled wis redd-oot
Frae the ill-makkit trauchles o men
His sang tyned awa frae hauch an frae wid
Whiles his shadda wis lowsed frae the ben.
Like some auld Hielan chief brocht tae plunner an grief
Tae be flushed frae ilk borrock or den
Nou they've re-biggit his fort an restored his estates
As Rid-gled's returned tae the glen.

Ilka burd in the lift noo jynes in the sang
Frae Gowdie tae wee Jenny Wren
It's been auld lang syne sin Rid-gled wis redd-oot
In years it's been hunner an ten
We'll tak up the cup an we'll drink tae's halth
Tae walcom him back hame again
An the laverock piper will skirl up a tune
Whan Rid-gled returns tae the glen.

THE BIG RID FLAMINGO

The Big Rid Flamingo has lang spirlie shanks
Circus stilt-walker o mudflats an banks
Skinnymalink neck, souple an lean
Bricht bonnie feathers wi cramasie sheen
Howks in the glaur wi its muckle strang neb
Tae find gustie morsels that dwall in the ebb
Like a sieve in its mou that is awmaist complete
It'll filter aff shrimps an sic braw things tae eat
For grace an yet gangliness
And pure lang-necked dangliness
The Big Rid Flamingo is gey haurd tae beat

Red Flamingos, White Flamingo, detail from full plate.

THE LANELY SLAVONIAN ON LOCH A CHOIRE

/ THE LONELY SLAVONIAN ON LOCH A CHOIRE

Doon at the RSPB car park the seasonal twitchers are oot
wi binos an telephotie lenses
traivelin the warld ower
fir a swatch at Loch Ruthven's Slavonian Grebes.
'This is the best place to see them in Britain.'
'There are several breeding pairs.'
'In America it's called the horned grebe.'

An so they'll heid for the bird hide
On Ruthven's birk clad banks
Tae catch a glisk o swimmin bird
the gowden ear tufts
bluid-rid ee like tiny bool, chestnut haunches
the weedy nest anchort in sedge.

Or mebbe lucky enough
Tae catch the coortship ritual
Dancin strictly breist-tae-breist
Ruthven rumba, the tangleweed tango
Twa flames astir oan the watter.

But for me A'll heid jist uphill tae Loch a Choire
Whaur the lanely Slavonian returns
Year on year tae the loch's ainly reed-bed
Like the hopeful romantic stood-up under the clock.

An so this spring A'll heid tae Loch a Choire again
In solidarity wi the grebe
In the hope that wan day …

Winter Falcon

THE FLYTIN O THE DOU AN THE GLED

(THE ARGUMENT BETWEEN THE DOVE AND THE HAWK)

The bonnie dou she taks the wing
An as she flees begins tae sing
I am the ane wha hirds the laund
The ane that feeds frae nature's haund
I luik doon on fields o grain
On wild wids an lands untamed
They mak a blessin when I pass
That the hairstin an the peace micht last
Atwixt the trees I glide above
Some cry me dou, some call me dove.

Higher up in the buinmaist air
The sun his targe in blinnin glare
I am the ane wha hings above
The ane that luiks doon on the dove
Wi ma cleuks I will scart her breast
An wi her feathers line my nest
Upon their flesh I've reared my young
Of hunting glories an joys I've sung
I am the scourge of every flock
Some cry me gled, some call me hawk.

Then drappin doon wi pynt tae pruive
Wi stentit clauts he made his muive
But gled misjudged his wistfu prey
Wha drappit law then winged away
Ye think me slaw ye think me weak
I am the ane ye ayeways seek
Come fly abune, come fly ablow
Through smore or lafty lift we'll go
And I'll lead you in a dance above
Come cry me dou, come call me dove.

Dou flew high then low again
She said gled did you nae ken
I am the burd o time untauld
They spak o me in legends auld
The een o love in Solomon's sang
I've stood ma grun but duin nae wrang
They coloured me as their whitest fleece
They made me into a symbol of peace
I am the burd of peace and love
Some cry me dou some call me dove.

He said Dou I will sieze my chance
For I read surrender in your olive branch
I kent men in days of auld
Wha prized me higher than finest gowd
My watchful ee wad keep them safe
When danger threatened at their gates
As aftentimes we'd hunt thegither
They treated me like feathered brither
Ower evry knowe an howe we'd stalk
Some cried me gled some called me hawk.

Dou said Gled if whit ye say is true
Then I'll plead that people mauna be like you
Wi nae predation tae be feared
Tae drouk this warld in sorrow'd tears
Gled said Dou guid luck wi that
But men still plunner when they're rich an fat
By means o force wi guile an stealth
They'll aim tae multiply their wealth
And this is how they raise their stock
Sae said the gled, sae told the hawk

Whan times are guid an winds are fair
They'll still gang reivin for the lion's share
My bonnie Dou did you nae ken
These are the weys o warldly men
For they are creatures of a different mind
Wha'll prey an herry eftir ther ain kind
Wher ther is plenty they will mak a dearth
Whan they are angry they will scorch the earth
An kill till ilk last drap is bled
Sae sang the hawk sae sang the gled.

They'll think it fine tae reap an sawe
The gruesome crop for the hoodie craw
For this is hou they come an go
These anes wha bide on laund below
Sae let's be happy as we mak this dance
For the joys o existence an oor circumstance
Let's birl an soar then turn again
Be thankfu we are burds not men
As aff intae the lift they sped
Sae flew the dou, sae flew the gled.

The Dove, The Red tailed Hawk, The Ash Coloured Hawk

Glenarbuck wid wis the buzzard
Liftin frae's tree
A rid flash an kestrel's eye view
Frae the tap o Haw Crags.

Boglairoch wis the peasies
Soarin ower tussocks
Wi a pee an a wee-eeet.
The Three Hills wis the laverock
Trillin in the first blue o clearin day.

The lonely tree on Duncombe crag
Wi craw's nest in calotype ootline
Like the frontispiece tae some folk legend
A secret story in the pages o the hills.

Lily Loch wis the douker's nest
The island o sedge wi'ts covered egg.
Crooks wis the short-eared owl in East Bay
The Hen Harrier rangin ower the glen.

The Gorge wis the skraich o peregrine
Shootin frae the banks like siller bullets
The burn itsel the slow wingbeat o heron
The whirr o dipper alang a daurk foam-swirlin pool.

Fieldfare an redwing closed summer
Jays flashed crimson-blue on banks
Heavy wi seasonal larder o red rowans.

Greylag wintered on lochside
Black-necked grebe foond some windblawn shelter
As in winter dimness ye wid wait
For peasie an curlew tae cry in spring again.

Snow Bunting, American Buzzard, Rusty Grackle, Bay winged Bunting

THE HAUCHTY MAVIS /

THE HOITY-TOITY THRUSH

Well don't I look awfully braw
With my speckled fur coat
Pecking at a croissant
Dropped at the delicatessen doorway
Warbling out a song in Charles Rennie MacIntosh letters
from the top of the tree
in Mrs McAllister's garden.
I have so much *smeddum*
so I do.

Wood Thrush, Red breasted Thrush/Robin, White breasted black-capped Nuthatch,
Red bellied black-capped Nuthatch

THE TEUCHIT / LAPWING

Witherty-wheep, wallopy-wheep
the teuchit's pipe gans skirlin
wallopy-wheep, witherty-wheep
abune the bog-rush pirlin

Whidderin through the caller blast
in snell an gowstie weather
ther's emerant sheen upon her wing
an bonnie creestit feather.

For ther's yowdendrift in the drumlie lift
and ice yet bords the linn
but ther's safter climes an blither times
at the back o the teuchit win.

Snipe

Red Starling, Female Starling, Warbler & Redfell

THE GLESGA STOOKIE / THE GLASGOW STARLING

The stookie taks tae Glesga skies
An quits the fields an hedges
Tae catch the latest blethers
On wires an windae ledges.
When Barraland Ballroom staurs cam oot
They'll meet in rowth o nummers
Tae haud ther raucous cooncil high
Up on the City Chaumers.
Amang the streets an moniments
It is easily jaloosed
Frae a splairge o guano graffiti
GLESGA STOOKIEZ ROOL THE ROOST

SOLAN / GANNET

The solan quits the open seas
Whan days are wearin lang
Tae white the monolithic Bass
A hunder an fifty thoosand strang.

They'll nest abune the riven ruins
O keep an martyrs prison
Tae flichter up on ghaistly prayer
And Covenanting vision.

A whirlin spindrift blizzard
They'll haunt the straits an narras
Tae drap taewards the briny deep
Like soarin spears an arras.

Raven, Vulture & Buzzard

YALLA-YITIE / YELLOWHAMMER

Baurley glowe
Gloamin fire
Flichterin spark
Frae roadside wire.

Black and Yellow Warbler, Autumnal Warbler

Orchard Orioles

Peerie-hawk & gleg hawk
Speugie & kay
Blue janet, swallae & merle
Firey-tail & Wullie-wagtail
Burnie becker
Blue Tam & Babblin Tam
Paitrick, Gourcock.

Kingfisher & Water Thrush, Tyrant Flycatcher, Great crested Flycatcher, Small green crested Flycatcher, Pewe (female), Wood Pewe (female)

OSPREY

Frae the banks an firths o Sierra Leone she comes
The fish hawk, the bald-gled
Across oceans, follaein shorelines an streams
Gien wide berth tae the Sahara
As ower days the wurld shape-shifts below
Mauritania, Morrocco
Portugal, Spain, France, England
Then becomes the naked bens
The daurk smudges o forest plantations
Till at last
The loch hoves intae view.

For six month o simmer licht
The loch will offer up its bounty
A troot below the ripple
Eyed frae a hunder an fifty fit
The jack-pike lurkin for fish or frog
Takken by sudden talon swoop.

They will bigg the nest in the tree-tap
Till wan late simmer's day the lone hunter
Broon-backit an white-briestit
Will be twa, three, fower
shapes high ower the watter.

Soon, Africa will call again.

Great Fish Hawk/Osprey, Snipe, Plover & Crow

WILSON'S BURDS /
WILSON'S BIRDS

As he humphed his pack through muirland nicht
Sawney Wilson heard the hummerin o the heather-bleat
like the chaunter
o some ghaist piper.

Or on the lanely heights
o Harthill an Shotts
luggin-in tae nocturnal skirl o the sea-pyot
as he shanked a guid sixty mile
frae Enbra tae Paisley.

Wilson on his wey hame frae some debate
frae Pantheon Haw filled wi thoosands
his heid thrang wi wirds
an ideals o Democracy.

Lowsed frae the strauchles o wabster
an rebel
his poems burnt on the Tolbooth steps
the Atlantic bore him tae a new warld.

Whaur he wid stravaig ten thoosand mile
wi fowlin-piece, pincil an brush
till the ornithology o America wis
in the grip o's haund.

A haill mixed flock wid be named eftir him.
Wilson's Storm Petrel.
Wilson's Plover.
Wilson's Phalarope.
Wilson's Warbler.
Wilson's Snipe.

Warbler & Canada Jay

MUCKLE SNIPPOCK / THE WOODCOCK

Of aw the burds tae east an west
The muckle snippock is the best.
Spreckelt an broon as the wild forest flair
Quately an glegly she'll wing through the air.
There is nane tae beat it, tis ae matchless sicht
The widcock abune when she's rodin at nicht.

Woodcock

MOCH HAWK / NIGHTJAR

The Moch Hawk haunts the rowthie wid
Tae wing wi the gloamin star
Chackin oot an antrin sang
An seal the nicht in a jar.

Male Whip poor will & female

SING LIKE A LINTY

Sing like a linty
Let your voice be heard
Each note an each cadence
An ilk spoken wurd.

Forget the detractors
For these are the ones
Wha stand on the sidelines
Like hunters wi guns.

They'll shoot down yer music
Ilk word that ye've sung
Tae pluck oot yer feathers
And tear out your tongue.

For here is the sport
And the game that they seek
Tae rile an tae herry
Ilk wurd that ye speak.

And even tho it's just plain wrang
They'll trash it 'chavvie'
'Rough'
or 'slang'.

So write doon yer language
Scrieve, let it fly
Sing like a linty
Ye're a burd in the sky.

Goldfinch, Blue Jay, Baltimore Bird

HAMISH MACDONALD

Hamish MacDonald is a poet, playwright and novelist with works published, broadcast and performed in Scots and English. He is author of the novel The Gravy Star and has written a number of plays which have toured throughout Scotland and abroad. He is a contributor to several publications by Scots language imprint Itchy Coo Publishing and has worked as writer and presenter on various BBC radio productions. He was the first Robert Burns Writing Fellow for Dumfries and Galloway Arts Association (2003-06) and the first Scots Scriever (2015-17), a post set up by Creative Scotland and the National Library of Scotland to raise awareness of the Scots language.